Assessment of Wearable Sensor Technologies for Biosurveillance

EXECUTIVE SUMMARY

Wearable sensor technology will shape the future of biosurveillance for the DoD, equipping the Warfighter with personal chemical, biological, and radiological exposure detection devices that can monitor both the environment around them, as well as the health of each individual service member. The Edgewood Chemical Biological Center (ECBC) has conducted a technology survey of the wearable sensor technology market for the Defense Threat Reduction Agency/Joint Science and Technology Office (DTRA/JSTO). Based on the information detailed in this report, the following actions are recommended:

- **The Department of Defense (DoD) should initiate an Advanced Technology Demonstration (ATD) for wearable technologies**. Innovative wearable technology candidates should be demonstrated to determine the extent of a device's capability, as well as its overall military utility.

- **DTRA/JSTO should create a Wearable Sensor Technology Research Consortium**. The development of biosurveillance-ready wearable sensor technologies will not be accomplished by a single academic, commercial, or government entity, but will require the coordinated efforts of multiple organizations. For success, common platforms should be selected by DTRA/JSTO and a biased-neutral test bed site developed in order to be successful. A wearable sensor technology research consortium, chaired by DTRA/ JSTO, should be formed to inform DoD leadership of current state of wearable sensor technologies so that long-range planning for the evaluation, integration and implementation of these technologies can be provided to CBDP stakeholders.

- **DTRA/JSTO should select a small number of common platforms to focus investment.** The DoD will not be the driver for technology development in the field of wearable sensor technology. Today, revenues for wearable sensor technologies are predicted to be more than **$59.7 billion by 2018**[1]. Identifying the most relevant technologies for advanced technology demonstration is highly advised, as existing commercial wearable sensors will likely need to be adapted by the DoD for biosurveillance use in the field.

- **JSTO/DTRA should engage FDA because medical decisions will likely result from wearable sensors**. Wearable sensor technology for biosurveillance missions will require access to bodily fluids. Passive wearable sensors alone will not be sufficient to conduct militarily relevant biosurveillance missions for targets of interest. In order to identify potential exposure to CBRN agents, semi-invasive technologies, such as the use of microneedles, will likely be required. The DoD should be aware that invasive sensors that access blood or are in close proximity to eyes will encounter apprehension and must overcome social taboos that may hamper wider acceptance.

- **Development of a biosurveillance-capable wearable sensor will require a multi-year investment.** Today, available commercial-off-the-shelf (COTS) wearable sensor technologies fall short of DoD mission needs. DTRA/JSTO should ensure that any proposed ATD for a wearable device outlines both realistic outcomes and timelines, and is programmed with the appropriate levels of RDT&E funding to support a 3 year developmental arc.

Considerations for Future of Wearable Sensor Technologies

Based on the above recommendations, DTRA/JSTO should be responsible for coordinating and driving investment into the development of biosurveillance-ready wearable sensors. ECBC has identified some additional considerations that DTRA/JSTO should be aware of during development:

- **Regulatory approval.** As the FDA approval process may take anywhere from two to ten years, many companies are reluctant to develop mobile technologies that draw medical conclusions or diagnoses. The FDA has proposed that certain mobile medical applications should be considered medical devices and be examined by the FDA prior to use. Wearable medical technology developers today generally side-step clinical validation costs by tracking general health metrics without drawing medical conclusions or diagnoses from the data collected. It is therefore critical to consider clinical validation costs when developing any new wearable medical technology.

- **Network compatibility.** Mobile wearable sensors will require an efficient and cost-effective method of transmitting data to centralized monitoring centers. Additionally, compliance with the DoD's information assurance policies and requirements may be required for devices that have the possibility of transmitting sensitive information.

- **Data Analysis.** The amount of data generated as a result of equipping every Warfighter with a wearable sensor could be unprecedented for the DoD. A significant effort will be required to ensure that the data that is collected is analyzed in an effective manner.

- **Personal data protection.** Equipping every Warfighter with a wearable sensor will require significant emphasis on securing the data collected as it may contain sensitive information about a service member's health. While electronic data capture and the use of digital signatures are quickly replacing error-prone traditional paper based medical transcription practices, consumer-generated health data from mobile monitoring systems, unlike information produced by doctors and hospitals, may not come with HIPPA protections.

- **Consider consulting with PEO-Soldier for guidance in wearable technologies.** PEO Soldier is responsible for developing, acquiring, fielding, and sustaining integrated equipment for Soldiers in a variety of operational environments. Early consultation with the PEO will help guide technologies early-on, ensuring that the equipment is not only suitable, but has a better chance of success and integration into operational systems.

TABLE OF CONTENTS

Introduction

Wearable sensor technology is fast revolutionizing the way in which we investigate and operate within the medical world. Today, real-time health information is no longer constrained to transmission via telephone, email or "snail mail", but can stream directly and instantly from portable monitoring devices to a wide variety of wireless mobile applications. From implantable wireless sensors that monitor arrhythmia and control heart rhythms, to "smart" contact lenses that measure ocular pressure or glucose levels in tears, today's wearable sensors enable continuous physiological monitoring with unprecedented portability and mobility while providing real-time access to patient information, faster diagnosis and detection, all at a fraction of the cost.

The global wireless health market is anticipated to more than double to $59.7 billion by 2018[1], allowing technological advancements in the field to become both readily available and adaptable to military purposes. By eliminating or replacing the somewhat cumbersome, conventional sensor equipment currently in use today, wearable sensors promise to modernize the field of medical and environmental monitoring while providing both a versatile and customizable capability to the Warfighter. By evaluating commercially available sensor technologies as well as those in development, decision makers can better-identify those wearable sensors that will best integrate into platforms targeted to enhance both Soldier protection and situational awareness.

Mobile biosensor platforms are quickly evolving to monitor a variety of their wearer's health metrics such as heart rate, EKG, temperature, pulse, respiration, and blood glucose. In the future, even more sophisticated mobile sensors could be developed that are capable of assessing bodily fluids for things such as stress levels, blood glucose, or even pathogen exposure (through the evaluation of things such as cytokines, oxygen, carbon dioxide and other metabolites). The measurements and assessment made by these devices will likely blur the lines between diagnostics and surveillance, with the main driver in this evolution being the smartphone which will create the standard interface that connects multiple devices together while making the almost instantaneous analysis and dissemination of gathered data possible.

Photo Removed Due to Copyright Restrictions

Compatibility with a mobile operating system (mobile OS) is imperative for any wearable device to be useful. Modern mobile OS must merge the elements of a personal computer (PC) with other components such as Bluetooth, Wi-FI, camera, touchscreen, speech recognition and GPS. While there are several mobile OS systems are available today, most systems are far behind the two biggest players: iOS and Android.

Device Compatibility Graph data from: Vandrico Solutions Inc. (http://vandrico.com)[6]

Photo Removed Due to Copyright Restrictions

Battery Life Histogram, Vandrico Solutions Inc.(http://vandrico.com)[6]

Battery life is of the upmost importance for wearable devices, especially when wearers do not have easy access to recharging capabilities. Device developers must consider and balance the need for long battery life with the inevitable size restrictions imposed on equipment used in the field. By building devices that leverage elements such as more efficient operating systems and lower power requirements, developers may be able to extend battery life, thereby addressing one of the most limiting factors known for these devices.

As wearable chemical and biological sensors become increasingly integrated into real-time interfaces, it becomes imperative for the Department of Defense (DoD) to identify and assess those sensor technologies that can be quickly developed and integrated into useable and deployable systems. By wrapping multiple diagnostic and monitoring modalities into a single device platform accessible through a smartphone[4], on-the-spot biomedical modeling, as well as physiological and cognitive monitoring can be immediately performed in order to improve the protection, performance, awareness and safety of special operators. Integrating, transmitting and displaying data in real-time will provide better reach-back support, give a higher level of situational awareness to incident commanders, and keep command control communications and intelligence (C3I) stakeholders better informed.

In the near term, we can expect that wearable innovations, whether funded by large corporate interests or small crowd sourcing efforts, will push the boundaries of how we perceive and utilize these technologies. The enormous level of competition - from both large and small developers alike – will likely deliver some form of wearable technology into the hands of nearly every citizen. Mid-term forecasts for this sector anticipate the development and approval of more precise medical diagnostics, behavioral software and fitness monitoring, while long-term predictions envision a future where humans, no longer encumbered by keyboard, mouse or monitor, will interact with computers much in the way they interact with other biological beings: by speech, gesture, or even expression.

As wearable devices become more ubiquitous and instrumental in science and data gathering, increasingly inexpensive wearable device technologies will replace their expensive and often cumbersome predecessors, making the real-time monitoring of health and environmental conditions for service members in both urban and remote areas more attainable than ever before.

Assessment of Wearable Sensor Technologies for Biosurveillance

The following report details wearable chemical and biological sensor technologies currently available or in development as of the summer of 2014. While a great amount of research has gone into the technology scanning for this report, (and given that funding for wearable technologies grew more than 80% in 2013 alone[2]), the devices listed are likely only the "tip of the iceberg" as to what technologies may be available within the next few years.

The technologies listed below fall into approximately three categories: 1) wearable medical sensors (such as watches or armbands), 2) "smart tattoos", and 3) environmental wearable sensors. Also included in this report are devices that fall outside the realm of biosurveillance sensors but may be relevant to the overall sensor platform, such as flexible battery technologies and flexible paper USB drives. The graphic below illustrates where many of today's available sensor devices can be worn.

Technology Summary

Medical Wearable Sensors			
Product	**Company**	**Picture**	**Description**
NUVANT MCT	Corventi		Continuous monitoring of symptomatic and asymptomatic cardiac abnormalities
Wearable Biological Sensor	Toshiba		Measures an electrocardiogram, pulse waves, body motions and skin temperature at the same time
"Smart" contact lens	Google/ Novartis		The lens can measure diabetics' blood sugar levels directly from tear fluid on the surface of the eyeball.
Modwells: Personal Modules for Wellness	Artefact Group		Input mods collect environmental and biometric data
Checklight	MC10 Inc.		Monitors for potential concussion
iWatch	Apple		Vitals such as blood pressure, heart rate, hydration level and other blood related data point (i.e., glucose)

Assessment of Wearable Sensor Technologies for Biosurveillance

Device	Maker	Image	Description
Smartwatch	Google	(mock up)	Heart rate, glucose, medication management
Smartwatch	Oxitone		Medical grade, wrist-worn pulse oximeter
In-ear Vitals sensor	Cis		Respiratory rate, pulse, heart rate, and oxygen content
EEG sensor	NeuroSky		Mobile, embedded research-grade EEG biosensor
Wearable vapor sensor	Genesis Nano Technology/University of Michigan		Continuous monitoring of diabetes, high blood pressure, lung disease and anemia.
Low-level electrical stimulator	Bioness®		Electrical stimulator for brain injury, stroke, nerve control

Vital Sign Monitor	Biovotion AG		Pulse oximetry and heat-rate monitoring
BodyMonitor	GEISS Institute for Social Intelligence – Germany		Wearable armband, measuring heart rate and skin conductance, and is used also to assess emotional state
Wearable Sensors	MIT Lincoln Laboratory/Bio engineering and Systems Technologies Group		Portable medical sensors and communications for health status and predictive care
Wearable biosensors	Joseph Wang (University California, San Diego)		Development of DNA and protein biosensors Non-invasive and minimally-invasive monitoring Remote sensors for environmental monitoring and security surveillance
Smart Contact Lens	Triggerfish		Measurement of internal eye pressure
Thin, flexible electronics	Wearable Computing		Transparent, flexible circuits small enough to wear on the surface of a contact lens

Assessment of Wearable Sensor Technologies for Biosurveillance

Flexible, integrated circuits	Ulsan National Institute of Science and Technology		Biosensors to monitor health conditions through tears
Vein detection glasses	Evena Medical		3D imaging of hard-to-locate veins, device interfaces with electronic medical records systems
Biometric smartwear	Omsignal		Captures breathing and heart recovery signals along with other biometric data, gives prescriptive notifications
Biometric smartwear	Hexoskin		Breathing rate, volume, cadence, ECG, sleep position, heart rate, and other physiological data
Wearable Wellnes System (WWS)	SMARTEX		Heart rate variability (simpato-vagal index), energy expenditure, body movement, posture (walking, lying, standing) and respiration
Smartphone mobile spirometer	MySpiroo		Portable peak expiratory flow meter
Behavior Analytics	Ginger.io		Identifies changes in behavior that may be warning signs for chronic issues like diabetes, depression, and cardiovascular disease.

Smart Tattoos and Skin Sensors

Product	Company	Picture	Description
Electronic "smart skin"	University of Illinois, John A. Rogers		Monitors vital signs
Tattoo Electronics	North-western University		Thermal sensors to monitor skin temperature and light detectors to analyze blood oxygen levels
Vibrating Smart Tattoos	Nokia		Magnetic tattoo that vibrates with your cell phone
BioStamp	MC10, Inc.		Senses temperature, heart rate, and other vital signs
Electrozyme	University of California, San Diego		Monitors health through analysis of a person's sweat
Tooth Tattoo	Princeton University and Tufts University		Detect oral disease by level of bacteria
Soft Microfluidic Assemblies	University of Illinois/ Shen Xu		Flexible assemblies of sensors, circuits, and radios for the skin

Environmental Wearable Sensors

Product	Company	Picture	Description
Mobile Geiger counter	Safecast		Mobile bGeigi Geiger counter that is designed to detect alpha, beta and gamma radiation
NODE Modular handheld sensors	Variable, Inc		Sensors are designed to detect motion, acceleration, CO_2, colors, temperature, air quality, barometric pressure, and ambient light
Portable air and water quality and sensors	Create Laboratories/ Carnegie Mellon		Monitoring of indoor fine particulate (PM2.5) and water purity
Lapka Environmental Monitor	Lapka, Inc.		Environmental sensors that plug into your iPhone and can detect radiation, electromagnetic feedback, nitrates in raw foods, and temperature and humidity
Flexible chemical sensors	Seoul National University		Polymer nanotubes that selectively interact with organophosphates
SensPods	Sensaris		Carbon dioxide(CO_2), Carbon monoxide (CO), Nitrogen oxide (NOx), Noise, Temperature, Humidity, Ozone (O_3), Luminosity (UVA, UVB, VC), A compact wireless PM measurement unit taking into account particles bigger than 1μ

Environmental Sensor	Sensorcon		H_2S and CO detection via Smartphone
Atmospheric pressure detector	Pressurenet		App that measures atmospheric pressure and provides measurements to better interpret weather related data. The app uses atmospheric sensors that are already in many Android phones.
Pocket molecular sensor	SCiO		Miniature spectroscope that detects unique optical signature, and thus determines the chemical composition of the material
Mobile electro-chemical detector	Harvard/ George M. Whitesides		A smart-phone enabled mobile electrochemical detector that tests for molecular-level health or environmental indicators

Thin-Flexible Batteries & Data Storage

Product	Company	Picture	Description
Zincpoly	Imprint Energy		Solid polymer electrolyte allows Imprint's batteries to be rechargeable
Cable Batteries	LG Chemical		Batteries with different shapes
Paper USB	intelliPaper		A "smart" paper business card bearing a detachable paper USB drive
Deep Tissue Power Generation	Stanford University/ John Ho		Wireless power transfer to deep tissue microimplants

Medical Wearable Sensors

Medical Wearable Sensors

Smart wearable medical devices are worn on the user's body (on places such as the wrist, skin or head) that offer features such vital sign, hydration, heart, blood glucose and other metabolic monitoring.

Corventi

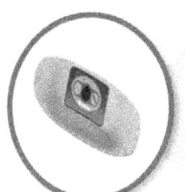

Platform: Nuvant Mobile Cardiac Telemetry System
Developer: Corventi
Use: Cardiac Monitoring
www.corventis.com

Designed to help physicians better diagnose and treat cardiac arrhythmias. The device provides continuous monitoring and recording of both symptomatic and asymptomatic cardiac irregularities by deriving and displaying ECG and heart rate, with periodical wireless data transmission. The MCT has received the CE mark as well as 501(k) clearance from the U.S. Food & Drug Administration (FDA).

Toshiba

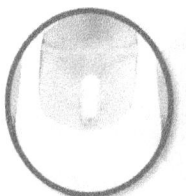

Platform: Wearable Vital Signs Sensor Module
Developer: Toshiba
Use: Cardiac Monitoring
http://www.semicon.toshiba.co.jp/eng/profile/news/newsrelease/topics_130305_e_1.html

Toshiba's "Silmlee Bar type," sensor is expected to be release in September of 2014. The sensor, which is not a pharmaceutical product at present, is designed for use by university, research institutes and other enterprises for evaluations and verification tests. The sensor, "Silmee Bar type," will be sold to universities, research institutes and enterprises. It is not a pharmaceutical product, and the company expects that it will be used for verification tests and evaluation. The Simlee© can calculate pulse waves, motion, skin temperature and electrocardio data via a small device that is worn on the chest using a gel pat. Results are displayed on a tablet or smartphone via Bluetooth.

Google

Platform: Smart Glasses & Smart Contact Lenses
Developer: Google
Use: Wearable health records, information sharing
http://www.google.com/glass/start

Google Glass

Google glass uses an optical head-mounted display (OHMD) to interact with the wearer like a smartphone, but in a hands-free environment. Private communications are achieved via built-in bone conduction speakers and a small screen (visible only to the user) that allows hands-free, private receipt of information while in the presence of others. The

technology opens up the possibility of using voice commands to record live-streaming videos of consultations and surgical operations, access digital health records, or take a picture that can be added to a patient's medical records.

Google's "smart" contact lens

Google is partnering with Alcon of Novartis to develop a smart contact lens whose low-powered microchip can measure blood sugar levels directly from tears and sends data to the wearer's mobile device to update blood sugar status. Also on the horizon are lenses that autofocus to improve long sightedness.

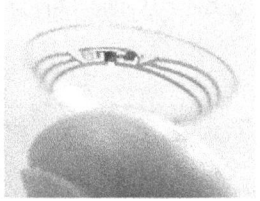

Modwells: Personal Modules for Wellness (Artefact Group)

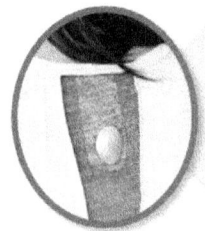

Platform: Personal Modules for Wellness
Developer: Modwells
Use: Collection of environmental and biometric data
http://www.artefactgroup.com/content/work/modwells-personal-modules-for-wellness/

Modwells "Mods" are a set of wearable health sensors that both collect and assess data that can be shared with healthcare professionals. The mods collect, assess and display both health and environmental data.

MC10, Inc.

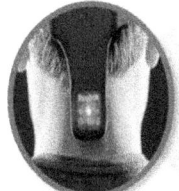

Platform: Checklight
Developer: MC10 Inc.
Use: Head impact monitor

http://www.mc10.com

MC10 is a company that specializes in wireless monitors that track degraded performance, hydration, temperature, and even traumatic brain injury. Checklight, Reeboks and MC10's flexible scull cap, was made to alert athletes when they may have received a dangerous blow to the head. The sensor is worn under athletic helmets and has a light built-in at the nape of the neck that glows when a potentially harmful blow has been received. MC10 also specializes in designing flexible displays, stretchable solar cells, and power-harvesting fabrics that can supply power to all of the wearer's electronic devices.

Apple Computer

Platform: iWatch
Developer: Apple
Use: Measurement of health related metrics

https://www.apple.com/ios/ios8/health/

Expected to be released sometime in late 2014, Apple's iWatch is anticipated to include 10 different health and fitness sensors that provide a general picture of health and fitness for the wearer. As an accompaniment device to the iPhone and iPad, the iWatch will measure various health-related metrics, with Apple's Health app letting users store data derived from a number of other fitness and health-tracking apps and devices while displaying this data on an easy to read dashboard. Vitals such as blood pressure, heart rate, hydration level and other blood related data point (i.e., glucose) will be monitored.

Google

Platform: SmartWatch
Developer: Google
Use: Heart monitor, health stats, medication management

http://www.google.com/glass/start/

Google smartwatch software is expected to be made available on Motorola, LG and Samsung watches. Google Fit's fitness tracking will display data such as heart rate, or detect whether its wearer has been physically active. Google's MediSafe medication management app allows users to take their medications both on time and safely.

Oxitone

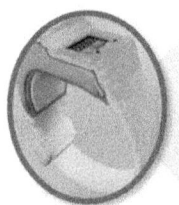

Platform: SmartWatch
Developer: Oxitone
Use: Medical grade, wrist-worn pulse oximeter

http://oxitone.com/

The Oxitone is a medical grade, wrist-worn pulse oximeter that continuously monitors heart rate and circumvents the need for the fingertip clip sensor and, oxygen saturation (SpO2), and respiration wirelessly from the wrist throughout the day. Capturing SpO_2 using photoplethysmography, the device offers real-time, non-invasive, remote two-way biofeedback.

CiS

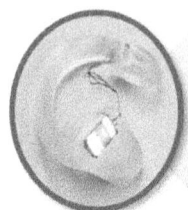

Platform: In-ear Vitals sensor
Developer: CiS
Use: Respiratory rate, pulse, heart rate, and oxygen content monitor

http://www.cismst.org/en/loesungen/im-ohr-sensor

CiS's research focuses on photovoltaics, microsystem technology and microsensors. Their in-ear cardiovascular monitoring sensor is based on an opto-electic transceiver module that allow continuous monitoring of vital signs such as respiratory rate, pulse, heart rate, and oxygen content. In-ear technology has greater stability, low orthostatic pressure modulation, stable temperature, and reduced tremor susceptibility The sensor works by detecting photon-tissue interactions via remitted light.

NeuroSky

Platform: EEG sensor
Developer: NeuroSky
Use: Mobile, embedded research-grade EEG biosensor
http://neurosky.com/products-markets/eeg-biosensors/

The NeuroSky Mindwave Electroencephalogram (EEG) Sensor uses a mobile, embedded research-grade EEG biosensor to amplify and processes raw brain signals, digitizing these analog electrical brainwaves to power user interfaces for games or research applications. Their interpretive algorithms and data analytics allow for a variety of mobile health applications, including their NeuroSky ECG biosensors which measures heart rate variability, heart rate recovery and respiration.

Genesis Nano Technology

Platform: Wearable vapor sensor
Developer: Genesis Nano Technology/University of Michigan
Use: Continuous monitoring of vitals &biomarkers of diseases
http://genesisnanotech.com/2014/08/u-michigan-researchers-develop-graphene-based-wearable-vapor-sensors/

The University of Michigan has developed a graphine-based wearable vapor sensor with the ability to continuously monitor diseases such as diabetes, high blood pressure, lung disease and anemia. The sensor works by detecting airborne chemical that are released through the skin or exhaled. Biomarkers of disease, such as acetone for diabetes or nitric oxide for the regulation of blood flow can also be monitored.

Bioness®

Platform: Low-level electrical stimulator
Developer: Bioness®
Use: Electrical stimulator for brain injury, stroke, nerve control

http://www.bioness.com/

The L300 Foot Drop System is an FDA-cleared medical device that is used as a rehabilitation tool for people with traumatic brain injury, cerebral palsy, spinal cord injury and stroke. The device uses Functional Electrical Stimulation (FES) to provide low-level electrical stimulation for muscles and nerve control.

Biovotion AG

Platform: Biovotion Vital Sign Monitoring (VSM)
Developer: Biovotion AG
Use: Pulse oximetry and hear-rate monitoring
http://www.biovotion.com/

The Vital Sign Monitoring (VSM) platform by Biovotion AG is a non-invasive mobile physiological monitoring device designed for people with chronic conditions. Optical sensors provide reliable pulse oximetry and hear-rate monitoring. The system data exchange and analysis is supported by integrated cloud service.

Body Monitor

Platform: BodyMonitor
Developer: GEISS Institute for Social Intelligence
Use: Wearable armband that measures vitals/emotional state

http://www.bodymonitor.de/

The BodyMonitor armband measures heart rate and skin conductance in order to assess the emotional well-being of the wearer. By using artificial intelligence (AI) and biometric data, the device captures electrodermal activity in real time to assess emotional states. Using the technique of signature analysis, the monitor is able to differentiate between both positive and negative valence of emotional attention response.

MIT's Lincoln Laboratory/Bioengineering and Systems Technologies Group

Platform: Portable medical sensors and communications for health status & predictive care in varying environments
Developer: MIT Lincoln Laboratory, Bioengineering and Systems Technologies Group
Use: Predictive performance and health monitoring

https://www.ll.mit.edu/mission/homeland/homelandprotectionsaccomplishments.html

The Bioengineering and Systems Technologies Group at MIT's Lincoln Laboratory is advancing the science of portable medical and physiological sensors, data processing algorithms and mobile applications. The sensors systems are geared towards preventing noise-induced hearing loss, avoiding heat casualties and musculoskeletal load injuries, and providing a variety of medical detection and treat modalities by monitoring physiological signals from ingested sensors from within the GI tract. Using both custom and commercial off the shelf (COTS) sensors, the laboratory's real-time physiological status monitoring (PSM) system features a series of custom tactical communications to be used in an on-body wireless network, and their "system on a chip" integrates sensing and communications into a single, small, low power device. Load sensors integrated into combat boots help prevent load injuries, wrist sensors take biomechanical measurements in the field, and helmet-mounted noise dosimeters help reduce noise-induced hearing loss. The lab's iBio test bed focuses on the integration of data gathered from molecular, genomic and life space monitoring tool to improve both health

IntegratedBioMedical System version 0.1.23

Individual	Traits	Activity	Sleep	Food	Rx/Exposures	Pathogens
Ancestry Exomes Genes Diseases Diseases:Genes	Heart Rate Summary Vocal .mat Pathways miRNA Clinical	Events		Meal	Drugs	

and overall performance. The group has collaborated with PEO Soldier, the U.S. Army Research Institute of Environmental Medicine (USARIEM), and the U.S. Army Natick Soldier RD&E Center (NSRDEC).

University of California San Diego

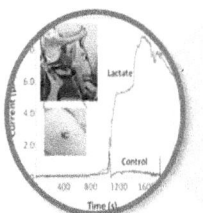

Platform: Wearable nano- biosensors
Developer: Joseph Wang (University California, San Diego)
Use: DNA and protein biosensors non-invasive and minimally-invasive monitoring

http://joewang.ucsd.edu/index.php?option=com_content&task=view&id=17&Itemid=35

Professor Wang's research activity is focused in the area of nanobioelectronic, a fast growing field directed at integrating electronic transducers with nano- and biomaterials. Work underway at Dr. Wang's laboratory include: textile-based wearable sensors, epidermal tattoos, DNA and protein sensors, forensic detection of explosives, remote environmental sensing, wearable power sources and smart tattoos. The laboratory's epidermal sensing devices conform to the wearer's anatomy can be used for non-invasive chemical monitoring. Their field-deployable hand-held micro analyzers are designed for environmental monitoring and clinical diagnostics as well as the evaluation of health status.

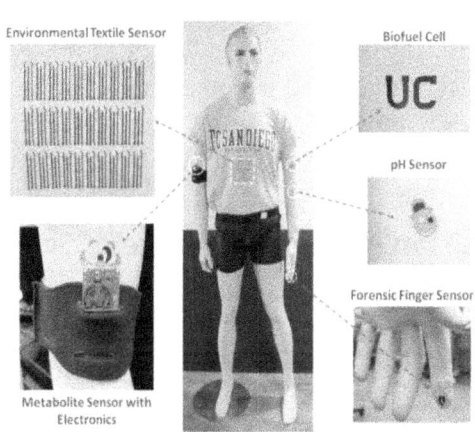

Triggerfish Smart Contact Lens

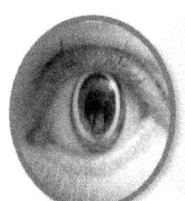

Platform: Smart Contact Lens
Developer: Triggerfish
Use: Measurement of internal eye pressure

http://health.ucsd.edu/news/2011/Pages/07-20-glaucoma-triggerfish-lens.aspx

In the U.S., large scale clinical trials have been launched to evaluate Tiggerfish's smart contact lens which measures internal eye pressure – a key risk factor for glaucoma, and a biomarker possible for chemical exposure. The lens provides long term monitoring of inner ocular pressure (IOP) and is made of a clear contact lens consisting of a ring-shaped strain gauge that monitors the shape of the cornea, a microprocessor and an antenna that can transmit pressure data to an external receiver. Data is transmitted by RF from the microprocessor to a receiver that records the data. Powered by an induction loop that generates small amounts of electricity from a magnetic field around the eye, the disposable device is intend for wear over a 24 hour period.

Wearable Computing Flexible Electronics

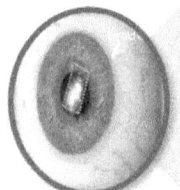

Platform: Thin, flexible electronics
Developer: Wearable Computing
Use: Small, transparent, flexible circuits

http://www.wearable.ethz.ch/research/groups/PlasticElectronics

Wearable Computing Group is developing a series of transparent, flexible circuits small enough to wear on the surface of a contact lens. The lenses are created by printing circuitry directly onto a layer of 1um thick layer of parylene which has been deposited on a vinyl polymer for support. The chip is then immersed in water, and the vinyl polymer is dissolved leaving only the thin circuitry. Afterward, the entire chip is placed in water, which dissolves the underlying polymer, leaving behind only the ultra thin layer of flexible circuitry about one-sixtieth the thickness of a human hair. Because of their thinness and biocompatibility, these films could be used for long-term medical monitoring.

Ulsan National Institute of Science and Technology's Flexible Integrated Circuits

Platform: Flexible, integrated circuits
Developer: Ulsan National Institute of Science and Technology
Use:

http://unist.storyzine.com/

Researchers at the Ulsan National Institute of Science and Technology have incorporated a light-emitting diode (LED) onto a commercial soft contact lens. Using a pliable mix of silver nanowires and graphine on the lens, scientist tested the lens in rabbits and found the electronics to continue functioning, even after five hours. The laboratory's eventual goal is to make wearable displays that rival Google Glass technology. By sandwiching silver nanowires in between graphine sheets they were able to make a composite material that stretched, had considerably lower electrical resistance (33 ohms/square), and transmitted 94 percent of visible light. By attaching the stretchy conductor to the contact lens and placing an LED on it, the researchers created a "primordial: display that may be an integral component in upcoming contact lens displays. The laboratory is currently investigating lenses that could serve as biosensors to monitor health conditions through tears or that filter light to counterbalance vision problems.

Evena Medical Vein Detection Glasses

Platform: Smart Glasses
Developer: Evena Medical
Use: 3D imaging of hard-to-locate veins

www.EvenMed.com

Evena Medical's Eyes-On Glasses System is a point-of-care wearable device that allows medical personnel to see "through" a patient's skin into the vasculature underneath. By utilizing wireless connectivity and 3D imagining, hard-to-locate veins can be found and accessed more easily, while the device interfaces with electronic medical records systems for on the spot documentation. The Epson Moverio "smart glasses" technology platform is one of the first healthcare applications for commercially available smart glasses.

Omnisignal

Platform: Biometric smartwear
Developer: Omsignal
Use: Biometric data gathering, prescriptive notifications

http://www.omsignal.com

Omsignal is the developer and manufacturer of biometric smartwear that streams and records real-time, continuous data wireless from the garment to a smartphone via Bluetooth L.E. Their smartwear captures breathing and heart recovery signals along with other biometric data, and even gives prescriptive notifications post-training to let the wearer know how their body is behaving over time.

Hexoskin

Platform: Biometric smartwear
Developer: Hexoskin
Use: Biometric data gathering

http://www.hexoskin.com/

Hexoskin is a smart shirt that includes three heart sensors that let the wearer track things such as breathing rate, volume, cadence, ECG, sleep position, heart rate, and other physiological data. A small device fits into a pocket at the waist of the shirt, allowing metrics are uploaded to an account and displayed on the wearer's smartphone.

SMARTEX

Platform: Wearable Wellnes System (WWS)
Developer: SMARTEX
Use: Heart rate variability, energy expenditure, body movement, and respiration

http://www.smartex.it/index.php/en/products/wearable-wellness-system

The SMARTEX Wearable Wellnes System (WWS) is designed to track physiological parameters while the wearer is moving. Data such as Heart rate variability (simpato-vagal index), energy expenditure, body movement, posture (walking, lying, and standing) and respiration are tracked and processed using signal processing algorithms. The software and processor assist in visualizing, managing and storing of data. Software and a powerful processor helps visualize, store and manage data. The ECG sensor is comprised of comprised of specialized yarns that are completely integrated into the structure of the garment.

MySpiroo Smartphone Spirometer

Platform: Cell phone-based spirometer
Developer: MySpiroo
Use: Portable peak expiratory flow meter

http://www.myspiroo.com

MySpiroo's smartphone mobile spirometer is designed to measure maximum voluntary ventilation (MVV), total lung capacity (TLC), Slow vital capacity (SVC), Functional Residual volume (RV), Forced expiratory volume (FEV), Forced vital capacity (FVC), peak expiratory flow (PEF), peak expiratory flow rate (PEFR), expiratory reserve volume (ERV), residual capacity (FRC) as well as reports and statistics and medication indexes. The portable peak flow meter connects to your smart phone, allowing wearers to both save their results and share them.

Ginger.io Behavior Analytics

Platform: Behavior Analytics
Developer: Ginger.io
Use: Behavioral change notifications

https://ginger.io/

A spin-off from the MIT Media Lab, Behavioral Health Analytics Company uses mobile phone "big data" to model user behavior to infer health and wellness. Changes in behavior may serve as warning signs for chronic issues like cardiovascular disease, depression and diabetes. If the system's algorithm suspects a change that may indicate a person is danger of a medical episode, it sends a message to a designated person (i.e. nurse, doctor or family member) with the idea that the communication will allow more timely intervention or prevention.

Proteus Digital Health

Platform: Ingestible Sensor
Developer: Proteus Digital Health
Use: Gathers data on medication taking, activity and rest patterns

http://proteusdigitalhealth.com/

Proteus Digital Health's FDA-approved ingestible sensor allows the user to better manage their medications. Ingested at the same time medication is taken, the device records when that medication was taken and sends the data to a patch worn on the skin. The patch, in turn, sends this data to a designated mobile device such as a smartphone or tablet.

Tissue-Integrated Sensors:

California Institute of Technology, Nanofabrication Group

Platform: Implantable medical sensors
Developer: California Institute of Technology, Nanofabrication Group/ Axel Scherer
Use: Glucose and ion sensor
http://nanofab.caltech.edu/research/microfluidics-biomedical.html

The nanofabrication group at Caltech is working on the development of implantable ion and glucose sensors that are integrated with communications systems and power sources to provide instantaneous medical data. With dimension of 100 um, these devices monitor the blood for real-time data gathering, making them well-suited for tasks such as drug administration or neural probes. Their skills in electronics, microfluidic systems and optical sensors are being directed towards the creation of miniature, inexpensive portable polymerase chain reaction (PCR) devices capable of performing disease diagnostics in remote locations.

Sandia National Laboratories/Ronen Polsky

Platform: Painless wearable microneedles
Developer: Sandia National Laboratories/Ronen Polsky
Use: Electrolyte monitoring
https://share.sandia.gov/news/resources/news_releases/electrolyte_sensor/#.U_9
C1nkrhkQ

Sandia is currently developing a wrist-worn diagnostic devices designed to continuously record and analyze low electrolyte levels. Real-time electrolyte analysis occurs through painless microneedles, so small that they are painless. The monitoring of other electrolytes such as sodium or calcium can also be adapted to the device by changing the selectivity of the carbon electrodes. The device was built to assess a soldier's health in the battlefield by continuously sampling interstitial fluid instead of blood, making it much less invasive than traditional sampling methods. In the future, the microneedles developed could also be used to administer drugs or electrolytes at the time of need.

Smart Tattoos

Smart Tattoos

The following technologies are smart tattoos that bind to the skin and incorporate electronic devices to monitor the human body. Smart tattoos are thin, stretchy sensors worn on the skin much like a temporary tattoo. They combine the comfort of a flexible sensor with electronic components, providing a human-machine interface that allows for medical monitoring and even pathogen detection.

University of Illinois

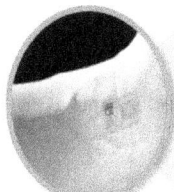

Platform: Electronic "smart skin"
Developer: University of Illinois, John A. Rogers
Use: Monitors vital signs

http://news.illinois.edu/news/11/0811skin_electronics_JohnRogers.html

Researchers at the University of Illinois have developed a temporary smart tattoo that allows for the monitoring of vital signs without the need for tape, conductive gel, electrodes or bulky wiring. Embedded in a film that is thinner than a human hair, these electronic sensors muse van der Waal force to allow the film to stick to skin without hindering motion, and are expected to stay on the wearer for up to two weeks. Currently, the sensors monitor temperature and heart rate, but scientist also believe these smart tattoos could also monitor brain wave, sense speech through the larynx, heat up to aid in wound healing, or even be designed to become touch sensitive , which could revolutionize the way in which artificial limbs are designed.

Northwestern University

Platform: Tattoo Electronics
Developer: North-western University
Use: Thermal sensors to monitor skin temperature and light detectors to analyze blood oxygen levels

http://www.northwestern.edu/newscenter/stories/2011/08/tattoo-electronics-huang.html

Northwestern University has designed a skin-mounted electronics with embedded circuitry in a flexible, bendable device. Discreetly worn and easily removed, the platform has been demonstrated with a wide variety of electronic components, including RF capacitors, solar power cells, transistors, LEDs, conductive coils and wireless antennas. Biomedical applications for these devices could include EEG and EMG, thermal sensors, blood oxygen sensors, and in situ EEG monitoring for understanding brain function outside of a laboratory. Through the use of electronic nanoribbons, the researcher conducted a speech reader experiment where they were able to able to differentiate between different vocabulary words, and control (with over 90 % accuracy) a voice-activated interface for a video game. This capability could revolutionize the way that patients with neurological or muscular disorders could interface with computer, and therefore the world around them.

Nokia Vibrating Tattoos

Platform: Vibrating Smart Tattoos
Developer: Nokia
Use: Magnetic tattoo that vibrates with your cellphone
http://conversations.nokia.com/2012/03/20/the-nokia-vibrating-tattoo/

Nokia has employed the use of ferromagnetic ink directly into a smart tattoo capable of alerting the wearer of things such as incoming calls or environmental alerts. The device works by generating a magnetic field that upon activating the ferromagnetic ink, produces perceivable and distinguishable vibration pattern in response to a signal (be it a text, voicemail or incoming call) that allows the wearer to differentiate between incoming messages.

MC10's Biostamp

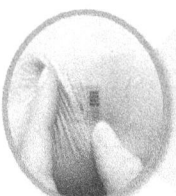

Platform: BioStamp
Developer: MC10, Inc.
Use: Senses temperature, heart rate, and other vital signs
http://www.mc10inc.com/

MC10's Biostamp prototype is a flexible, wearable computing device that provides a collection of sensors that can be applied to the skin in the same way as a temporary tattoo. Using near field communication (a wireless technology such as that used in E-Zpass) the sensors in the device collect data such as heart rate, brain activity, temperature and UV exposure while sending the data to a smartphone for further analysis. The Biostamp could be worn for about two weeks, providing full- time medical monitoring, a capability that could forever change the way in which medical diagnosis are made.

Elecctrozyme's smart tattoo

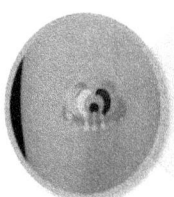

Platform: Electrozyme
Developer: University of California, San Diego
Use: Monitors healthy through analysis of person's sweat
http://electrozyme.com/

Electrozyme's electrochemical biosensor tattoo was originally developed by researchers at the Laboratory for Nanobioelectronics at UCSD. The printed sensor analyses in real-time chemical elements in the wearer's sweat and evaluates things such as lactate buildup to quantify muscular exertion, electrolyte balance, and even epidermal pH to assess hydration levels. Proprietary algorithms recognize and integrate physiology patterns from the sensors in real time. In addition to their biosensors, Electrozyme is also developing an epidermal biofuel cell that they hope will power their wearable sensor by capturing the biochemical fuels found in sweat.

Princeton and Tufts University

Platform: Wireless Tooth Tattoo
Developer: Princeton and Tufts University
Use: Magnetic tattoo that vibrates with your cell phone

http://www.princeton.edu/main/news/archive/S33/79/62E42/index.xml?section=science

Princeton and Tufts University scientist are developing an electronic wireless tooth tattoo designed to measure bacterial levels in the mouth in order to help in the detection of gum disease. The sensor, made of silk, graphite and gold, is much less invasive than drawing blood and promises to one day detect more than just periodontal disease by assessing other indicators of disease that may also be measure from saliva. The scientists are looking increase the specificity of the sensor by constructing peptides that have the ability to bind with specific strains of bacteria.

University of Illinois at Urbana-Champaign/ Shen Xu

Platform: Soft microfluidic assemblies

Developer: University of Illinois at Urbana-Champaign/Sheng Xu,

Use: "Stick-on" electronics

https://www.sciencemag.org/content/344/6179/70?related-urls=yes&legid=sci;344/6179/70

Dr. Xu and his colleagues at University of Illinois are experimenting with controlled mechanical buckling, soft microfluidics, and structured adhesive surfaces to develop skin-mounted power supplies, circuits, sensors and radios geared towards the development of wireless, clinical grade physiological monitoring.

Environmental Wearable Sensors

Environmental Wearable Sensors

These small, portable sensors monitor environmental conditions to provide early warning for exposure or assessment of contamination. Many of these sensors work in combination with a smartphone, transmitting data quickly and wirelessly.

Safecast

Platform: Mobile Geiger counter
Developer: Safecast
Use: Mobile Geiger counter that detects alpha, beta and gamma radiation

https://www.kickstarter.com/projects/seanbonner/safecast-x-kickstarter-geiger-counter

Driven by events following the Fukushima catastrophe in Japan, Safecast has developed the mobile bGeigi Geiger counter that is designed to detect alpha, beta and gamma radiation. The sensor was designed in order to form a sensor network that could detect and map nuclear radiation so that government organization as well as the general public could help provide a more accurate and detailed information about radiation contamination. Available on Amazon.com as a kit, design is also completely open source to make the technology available to everyone (see: http://www.bunniestudios.com/blog/?p=2218).

Variable, Inc.

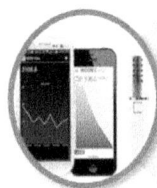

Platform: NODE Modular handheld sensors
Developer: Variable, Inc
Use: Sensors that detect motion, acceleration, CO_2, colors, temperature, air quality, barometric pressure, and ambient light

http://variableinc.com/

Variable, Inc.'s NODE's modular handheld sensors are designed to detect motion, acceleration, CO_2, colors, temperature, air quality, barometric pressure, and ambient light. The platform, which uses sensor modules that are interchangeable, also has a wireless barcode reader, a magnetometer, accelerometer and a gyroscope.

Create Laboratories

Platform: Portable air and water quality and sensors
Developer: Create Laboratories/Carnegie Mellon
Use: Monitoring of indoor fine particulate (PM2.5) and water purity

http://www.cmucreatelab.org/projects/Speck

The Community Robotics, Education and Technology Empowerment Lab (CREATE Lab) out of Carnegie Mellon are the developers of portable air quality and water quality sensors designed for personal exposure monitoring. Their Speck indoor fine particulate (PM2.5) monitor monitors pollutants in the air at 2.5 micrometers in diameter or smaller. The monitor is small and portable and connects via USB to a power connector. Create Laboratory's WaterBod continuously collects data and uploads water purity information to the internet through a ZigBee-installed module.

Lapka, Inc.

Platform: Environmental Monitor
Developer: Lapka, Inc
Use: Mobile sensors that detects radiation, electromagnetic feedback, nitrates in raw foods, temperature and humidity

http://about.mylapka.com/pem/product/manual/

Lapka, Inc has created a set of iPhone-readable sensor capable of detecting nitrates in raw food, temperature, humidity, electromagnetic feedback, breath alcohol, and even radiation. leveraging the battery and processing power of the iPhone, Lapka's sensors can remain small. Each measurement taken by a sensor is converted to a block on the home screen of the app and displays the symbol for individual sensor used to take the measurement. With each passing day, more home screens are added to the app, allowing the user to intuitively scan through readings and trend data. Entries are uploaded to the cloud and shareable.

Seoul National University

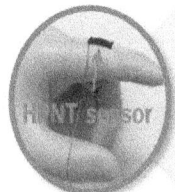

Platform: Flexible chemical sensors
Developer: Seoul National University
Use: Polymer nanotubes that selectively interact with organophosphates

http://pubs.acs.org/doi/pdf/10.1021/nl204547t

Researchers at the School of Chemical and Biological Engineering at Seoul National University have developed a flexible chemical sensor based on multidimensionally hydroxylated poly(3,4-ethylenedioxythiophene) (PEDOT) nanotubes (HPNTs) that can detect the sarin stimulant methylphosphonate (DMMP) at a detection limit of 10 ppt. The sensor material is connected to an electrical power source and display, and developers plan to use this technology to create a wearable nerve gas detection sensor that will replace the ion mobility spectrometer devices worn by soldiers and security personnel. The group's polymer nanotubes, which selectively interact with organophosphates, have a large surface area that allows them better interaction with chemical vapors, making them and have are easy to scale-up making them, according to Jyongsik Jang, a polymer scientist, 100–1000 times more sensitive than other existing DMMP sensors. While still in early development, these polymers show real promise towards the development of a soft, wearable chemical agent detector.

Sensaris

Platform: SensPods
Developer: Sensaris
Use: Gas, noise, humidity, temperature, luminosity and particulate matter sensor
http://www.sensaris.com/

Sensaris is a company that specializes in the development of wireless, low powered sensors specifically designed to be compatible with mobile applications. Sensaris' SensPods are wireless, compact sensors that measures things such as temperature, carbon monoxide (CO), Nitrogen oxide (NOx), humidity, Ozone (O_3), Luminosity (UVA, UVB, UVC), noise, and particles bigger than 1μ. All SensPods come equipped with and Android application and have access to SensDot, their web interface.

Sensorcon

Platform: Environmental Sensor
Developer: Sensorcon
Use: H$_2$S and CO detection via Smartphone

http://sensorcon.com/

Sensordrone is a mobile portable gas detector. The sensor, which pairs with your smartphone, can serve as a lux meter, a weather station or thermometer, and senses humidity and gases such as H$_2$S and CO by using specific apps. The device is small enough to fit on a keychain and communicates with a smartphone via Bluetooth.

PressureNet

Platform: Atmospheric pressure detector
Developer: Pressurenet
Use: The app leverages atmospheric sensors that are already in many Android phones

https://pressurenet.io

Pressurenet is an atmospheric pressure detector app that connects smartphones to a high resolution weather data platform. The Android-powered app measures atmospheric pressure by using sensors that are already in most Android phones. Data is shuttled to a website where is can aid in making weather predictions or evaluating the effects of atmospheric on environmental systems.

SCiO Pocket Sensor

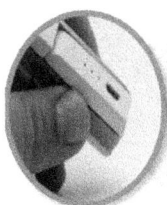

Platform: Pocket molecular sensor
Developer: SCiO
Use: Miniature spectroscope that detects unique optical signature for chemical composition

https://www.kickstarter.com/projects/903107259/scio-your-sixth-sense-a-pocket-molecular-sensor-fo

The SCiO pocket sized molecular sensor scans material or physical objects to determine the chemical composition of a material. A miniature spectroscope, it working by shining near-infrared light on the object being scanned, thereby exiting their molecules to give a unique optical signature that helps determine the chemical make-up of the material. Results are sent to the cloud, where data is processed in real-time by algorithms and results are posted to the user's phone within seconds. The SCiOs integrated battery provides approximately one week of use for every charge

Harvard/George Whitesides

Platform: uMED
Developer: Harvard/ George Whitesides
Use: A smart phone-enabled mobile electrochemical detector that test for molecular level health or environmental indicators
http://www.pnas.org/content/111/33/11984

The uMED device analyzes samples via differential pulse voltammetry, chronoamperometry, potentiometry, square wave voltammetry and cyclic voltammetry. The sensor has been shown to perform an electrochemical enzyme-linked immunosorbent assays (ELISAs) for the detection of malaria antigen, monitor levels of sodium in urine samples, measure blood glucose, and even detect trace amounts of heavy metals in water. Avoiding the need to enter point-of-care data by hand, the device transfers data using a voice-based cell phone approach. The device also provides on-board sample mixing and is expected to cost approximately $25.

Thin-Flexible Batteries and Data Storage

Thin-Flexible Batteries and Data Storage

Critical to the development of wearable technologies is the need for power and data. Enabling technologies such as data analysis and storage, as well as increased batter life will be key to the success of these devices.

Imprint Energy

Platform: Zincpoly™
Developer: Imprint Energy
Use: Solid polymer electrolyte allows Imprint's batteries to be rechargeable

http://www.imprintenergy.com/

Imprint Energy's Zincpoly™ technology is based on thin flexible rechargeable zinc batteries that are less toxic than traditional lithium ion batteries. Their low toxicity, which makes them more suitable for medical implants, along with their rechargeability and lower production costs makes them particularly suited for wearable devices.

LG Chemical

Platform: Cable Batteries
Developer: LG Chemical
Use: Flexible batteries

http://www.lgcpi.com/chem.shtml

LG Chemical is advancing battery technologies that produce batteries with different shapes. These battery types include: step batteries, which allow stacking of two or more batteries that adapting to devices of differing shapes; curved batteries, which are more suitable for curved devices such as watch bands; and cable batteries, which are waterproof, bendable batteries whose low electricity prevents them from overheating – making them very suitable not only for wearable devices, but also possibly for some of the newer, up and coming smart fabric technologies.

intelliPaper's Paper USB

intelliPaper is a "smart" paper business card that sports a detachable paper USB drive.

Platform: Paper USB
Developer: intelliPaper
Use: Smart paper business card with detachable paper USB drives

https://www.intellipaper.info/business-cards/

Individual cCards can be updated remotely with a back-end interface that allows the user to log into their account and make changes to the data. The card can be tracked and will indicate whether the recipient has opened it.

Stanford University/ John S. Ho

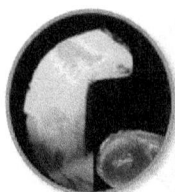

Platform: Power Generation
Developer: Stanford University/ John Ho
Use: Wireless power transfer to deep tissue microimplants

http://www.pnas.org/content/early/2014/05/14/1403002111.abstract

Stanford University is developing methods for the wireless generation power to deep tissue microimplants[5]. In this system, the research use midfield powering to create areas of high energy density deep within tissue to power a 2mm microimplant, a device exponentially smaller than a conventional pacemaker.

REFERENCES

1. marketsandmarkets.com, *Wireless Health Market (WLAN, WMAN, WPAN, Sensors, Smartphone's, Tablet PC, Mobile APPS) - Global Trends, Opportunities, Competitive Landscape & Forecasts (2013 - 2018)*, Publishing Date: September 2013, Report Code: HIT 2102. http://www.marketsandmarkets.com/Market-Reports/wireless-healthcare-market-551.html

2. CB Insights, http://www.cbinsights.com/blog/wearable-tech-venture-capital/

3. FDA: Mobile Medical Applications, http://www.fda.gov/MedicalDevices/ProductsandMedicalProcedures/ConnectedHealth/MobileMedicalApplications/ucm255978.htm

4. Awgatty B, et al, *Fluorescent sensors for the basic metabolic panel enable measurement with a smart phone device over the physiological range*, Analyst, 2014 Aug 15., (Epub ahead of print: http://www.ncbi.nlm.nih.gov/pubmed/25126649).

5. J. S. Ho, et al, *Wireless power transfer to deep-tissue microimplants*, PNAS, 111, 7974-7979 (2014).

6. Wearable Market Insights (1st Quarter, 2014), Vandrico Solutions Inc., 288 East 1st Street, North Vancouver, BC Canada V7L 1B3, info@vandrico.com, (604) 229-1215.

WEBSITES

Medical Wearable Sensors

1. **Corventi**
   www.corventis.com
2. **Toshiba**
   http://www.semicon.toshiba.co.jp/eng/profile/news/newsrelease/topics_130305_e_1.html
3. **Google**
   http://www.google.com/glass/start
4. **Modwells**
   http://www.artefactgroup.com/content/work/modwells-personal-modules-for-wellness/
5. **MC10, Inc**
   http://www.mc10.com
6. **Apple's iWatch**
   https://www.apple.com/ios/ios8/health/
7. **Googles's Smart Watch**
   http://www.google.com/glass/start/
8. **Oxitone**
   http://oxitone.com/
9. **CiS**
   http://www.cismst.org/en/loesungen/im-ohr-sensor
10. **NeuroSky**
    http://neurosky.com/products-markets/eeg-biosensors/
11. **Genesis Nano Technology**
    http://genesisnanotech.com/2014/08/u-michigan-researchers-develop-graphene-based-wearable-vapor-sensors/
12. **Bioness®**
    http://www.bioness.com/
13. **Biovation AG**
    http://www.biovotion.com/sensors.phtml
14. **Body Monitor**
    http://www.bodymonitor.de/   (site is in German)
15. **MIT's Lincoln Laboratory/Bioengineering and Systems Technologies Group**
    https://www.ll.mit.edu/mission/homeland/homelandprotectionsaccomplishments.html
16. **University of California San Diego**
    http://joewang.ucsd.edu/index.php?option=com_content&task=view&id=17&Itemid=35
17. **Triggerfish Smart Contact Lens**
    http://health.ucsd.edu/news/2011/Pages/07-20-glaucoma-triggerfish-lens.aspx
18. **Wearable Computing Flexible Electronics**
    http://www.wearable.ethz.ch/research/groups/PlasticElectronics
19. **Ulsan National Institute of Science and Technology's Flexible Integrated Circuits**
    http://unist.storyzine.com/

20. **Evena Medical Vein Detection Glasses**
    www.EvenMed.com
21. **Omnisignal**
    http://www.omsignal.com
22. Hexoskin
    http://www.hexoskin.com/
23. **SMARTEX**
    http://www.smartex.it/index.php/en/products/wearable-wellness-system
24. **MySpiroo Smartphone Spirometer**
    http://www.myspiroo.com/#intro
25. **Ginger.io Behavior Analytics**
    https://ginger.io/
26. **Proteus Digital Health**
    http://proteusdigitalhealth.com
27. **California Institute of Technology Nanofabrication Group**
    http://nanofab.caltech.edu/research/microfluidics-biomedical.html
28. **Sandia National Laboratories/Ronen Polsky**
    https://share.sandia.gov/news/resources/news_releases/electrolyte_sensor/#.U 9C1nkrhkQ

Smart Tattoos

29. **University of Illinois**
    http://news.illinois.edu/news/11/0811skin_electronics_JohnRogers.html
30. **Northwestern University**
    http://www.northwestern.edu/newscenter/stories/2011/08/tattoo-electronics-huang.html
31. **Nokia Vibrating Tattoos**
    http://conversations.nokia.com/2012/03/20/the-nokia-vibrating-tattoo/
32. **MC10's Biostamp**
    http://www.mc10inc.com/
33. **Elecctrozyme's smart tattoo**
    http://electrozyme.com/
34. **Wireless Tooth Tattoo/Princeton and Tufts University**
    http://www.princeton.edu/main/news/archive/S33/79/62E42/index.xml?section=science
35. **University of Illinois at Urbana-Champaign/ Shen Xu**
    https://www.scincemag.org/content/344/6179/70?related-urls=yes&legid=sci;344/6179/70

Environmental Wearable Sensors

36. **Safecast**
    https://www.kickstarter.com/projects/seanbonner/safecast-x-kickstarter-geiger-counter
37. **Variable, Inc.**
    http://variableinc.com/

38. **Create Laboratories**
    http://www.cmucreatelab.org/projects/Speck
39. **Lapka, Inc.**
    http://about.mylapka.com/pem/product/manual/
40. **Seoul National University**
    http://pubs.acs.org/doi/pdf/10.1021/nl204547t
41. **Sensaris**
    http://www.sensaris.com/
42. **Sensorcon**
    http://sensorcon.com/
43. **PressureNet**
    https://pressurenet.io/
44. **SCiO Pocket Sensor**
    https://www.kickstarter.com/projects/903107259/scio-your-sixth-sense-a-pocket-molecular-sensor-fo
45. **Harvard/George Whitesides**
    https://www.pnas.org/content/111/33/11984

Thin-Flexible Batteries and Data Storage

46. **Imprint Energy**
    http://www.imprintenergy.com/
47. **LG Chemical**
    http://www.lgcpi.com/chem.shtml
48. **intelliPaper's Paper USB**
    https://www.intellipaper.info/business-cards/
49. **Stanford University/ John Ho**
    http://www.pnas.org/content/early/2014/05/14/1403002111.abstract

David L. Hirschberg, PhD

Assistant Professor, Clinical Pathology
Columbia University and Chief Technology
Officer at the Center for Infection and
Immunity

Kelley Betts, MS

Leidos, Inc.
U.S. Army Edgewood Chemical Biological Center BioSciences Division
5183 Blackhawk Road
Aberdeen Proving Ground, MD 21010

Peter Emanuel, PhD

JUPITR ATD Lead/ECBC BioSciences Division Chief
5183 Blackhawk Road
Aberdeen Proving Ground, MD 21010

Matt Caples, PhD

Science Advisor to BioSciences Division
5183 Blackhawk Road
Aberdeen Proving Ground, MD 21010

The recommendations detailed in this document were based on the assessments of the current state of the art technologies for wearable sensor technologies and on how these technologies are both maturing and evolving.

END

www.ingramcontent.com/pod-product-compliance
Lightning Source LLC
Chambersburg PA
CBHW080615180526
45168CB00007B/2925